W9-COT-222

BANANAS

A TRUE BOOK

by

Elaine Landau

Children's Press®
A Division of Grolier Publishing
New York London Hong Kong Sydney
Danbury, Connecticut

A banana sailboat

Reading Consultant
Linda Cornwell
Coordinator of School Quality and Professional Improvement, Indiana State Teachers Association

Author's Dedication
For Nicole Garmizo

Visit Children's Press® on the Internet at: http://publishing.grolier.com

Library of Congress Cataloging-in-Publication Data

Landau, Elaine.
Bananas / Elaine Landau.
p. cm. — (A True book)
Includes bibliographical references (p.) and index.
Summary: Examines the history, cultivation, and uses of bananas.
ISBN 0-516-21025-4 (lib. bdg.) 0-516-26574-1 (pbk.)
1. Bananas—Juvenile literature. [1. Bananas.] I. Title. II. Series.
SB379.B2L35 1999
634'.772—dc21 98-47328
CIP
AC

A Division of Grolier Publishing Co., Inc.

Printed in the United States of America.
1 2 3 4 5 6 7 8 9 10 R 08 07 06 05 04 03 02 01 00 99

Contents

The Appealing Banana 5
A Slice of Banana History 9
All About Bananas 20
Bananas—A Healthy Treat 31
Choosing Bananas 36
Banana Trivia 40
To Find Out More 44
Important Words 46
Index 47
Meet the Author 48

People enjoy bananas
at every meal.

The Appealing Banana

Bananas have long been a favorite food of millions of people around the world. They are the world's best-selling fruit. In the United States alone, more than eleven billion bananas are eaten each year.

Bananas are often enjoyed as snacks. Many people slice

Banana slices make a bowl of cereal extra nutritious (above). The banana split was invented in 1920 (right).

them into cereals. And they are also found in fruit salads.

Frequently this tasty fruit is used in desserts as well. There's banana cake, banana milk shakes, banana cream pie, frozen banana pops, and banana pudding. And don't forget the king of all ice cream sundaes—the banana split or banana royal.

Some people in warm areas of the United States have banana plants in their backyards. But bananas are not grown commercially (for sale to stores and other places) anywhere in the country. All the bananas bought at U.S. supermarkets are shipped from other places.

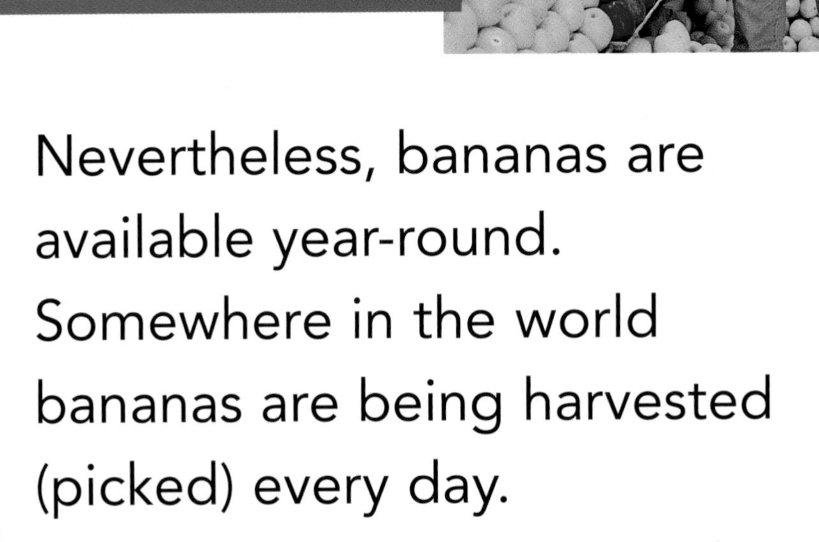

Bananas make a long journey from tropical plantations to our supermarkets.

Nevertheless, bananas are available year-round. Somewhere in the world bananas are being harvested (picked) every day.

A Slice of Banana History

People have been eating bananas for thousands of years. Scientists who study plants think this fruit first grew in Asia. The earliest bananas might have come from the hot, steamy jungles of Malaysia. But those

People in ancient times transported bananas.

bananas weren’t like the ones we have today. They were probably tough and filled with seeds.

The banana may have been among the first fruits farmed by people. When the conqueror

Alexander the Great and his armies entered India in 327 B.C.— they found bananas there. We don't know how long bananas had been grown in the region before that. But wise men in India often sat

Alexander the Great meets a ruler in India on the battlefield.

beneath banana trees and ate the fruit. Therefore, it was thought that bananas made you very smart.

In time, the banana spread to many different parts of the world. As people traveled to trade or conquer, they often took the roots of the plant with them. Arab traders brought the banana plant across the Indian Ocean to the tropics of Africa's east coast. The Africans liked the fruit. And before long, bananas were being planted on other parts of the continent.

In 1482, Portuguese explorers arriving in Africa enjoyed the fruit as well. They later brought the banana to their colonies in the Canary Islands. But before long the Canary Islands were taken over by Spain. And in 1516 it was a Spanish friar who took the banana plant with him to the Americas.

The friar had gone to the Caribbean island of Hispaniola to introduce Christianity to the Indians there. However, besides planting the seeds of his religion, he also planted the

The banana tree (far right) and other fruit trees of Hispaniola

roots of several banana plants. The banana plants grew well in the tropical island's rich, fertile soil. But this was not the end of the banana's travels. In time, it was also planted in Mexico, Paraguay, Peru, and other areas.

During the 1800s, bananas were sometimes brought to the United States by sailors. These men had been sailing in the Caribbean when they came across the fruit. However, bananas made their official debut

Bananas started appearing on New York City docks in the 1870s (left). They were officially introduced in 1876 at the Philadelphia Centennial Exhibition (below).

in the United States in 1876 at the Philadelphia Centennial Exhibition. These bananas were individually wrapped in foil and sold for fifty cents each.

Today large numbers of bananas are grown in Central America, the West Indies, and regions of South America. Banana plants do well in hot, moist climates. About 90 percent of the bananas eaten in

Young banana plants on a plantation in the French West Indies

the United States come from Latin America. The fruit is also heavily planted in Africa, the East Indies, and the Pacific Islands.

Though mostly grown in the tropics, some bananas also come from Iceland! These bananas are cultivated (grown) in special hothouses. Underground volcanic springs heat the water used to keep these enclosed areas quite warm. Some of the bananas

This greenhouse in Iceland (above) is heated by steam from a hot spring (right) nearby.

grown in Iceland are eaten there. But others are exported (shipped out) to other countries.

All About Bananas

Do you think bananas grow on trees? This isn't quite true. Banana plants look like trees because they reach a height of between 8 and 40 feet (2.4 and 12 meters). But they aren't trees because they don't have woody trunks or branches.

Bananas and lilies (below) share family ties.

Instead they are large plants. Banana plants are actually giant herbs. They are members of the same family as lilies and orchids. Banana plants are among the largest plants on earth without woody stems.

The pseudostem of a banana plant

The banana plant's trunk is made up of tightly wrapped, overlapping leaves. The trunk is also known as the pseudostem (SOO-doh-stem). A banana plant's pseudostem can be quite broad—sometimes as wide as 2 feet (0.6 m).

A cluster of blossoms (left) will later become miniature banana "fingers" (right).

As the banana plant matures, a stem with a large bud begins to grow inside it. Soon this stem will poke its way through the trunk. The purple petal-like leaves of the bud open to reveal tiny blossoms.

These flowers develop into very small green bananas. Each group or cluster of these bananas is called a hand. Not surprisingly, the individual bananas are called fingers.

A banana-plant hand has far more fingers than a human hand. Between 14 and 20 fingers (or bananas) grow on each hand. And usually banana plants have 7 to 9 hands. Therefore, a banana plant may yield somewhere between 120 and 180 bananas.

Once the stem is heavy with fruit (right), it is ready for harvesting.

"Banana bags" battle bugs!

As the fruit grows, farmers sometimes cover the banana plant's stem with a large plastic bag. This protects the fruit from strong winds and harmful insects. But it still allows the sunlight in.

Once a banana has ripened, it will fall off the plant. After this occurs it must be eaten soon afterward. If not, it will rot.

Bananas sent to distant areas are handled differently. These are plucked from the plant while they are still green. Then they are washed and checked for

Before packing and shipping, bananas are cleaned in large water tanks.

bruises. The bananas are packed in boxes to protect them during shipping. About 40 pounds (18 kilograms) of bananas fit into each box.

The boxes are loaded into refrigerated containers for the trip. The temperature remains at about 57 degrees Farenheit (14 degrees Celsius) inside the containers. This will stop the fruit from ripening too soon.

Once they arrive, the bananas are put in special

A ripening room

temperature-controlled ripening rooms. Here high humidity and natural gases are used to ripen the fruit. As a banana ripens, the fruit's starches turn to sugar. The riper the banana, the sweeter it is.

A farmer's market

After a few days in the ripening rooms, the green bananas will turn yellow with brown markings. They'll look and taste just as if they had ripened on the plant. And soon afterward, you'll find them at your grocery store.

Bananas—A Healthy Treat

Bananas are good for you! They are high in Vitamins B6, C, and A. Bananas are also rich in potassium. Potassium is necessary for normal growth. And it helps to keep your skin healthy.

In addition, eating bananas provides you with other valuable

The banana's yellow skin keeps the fruit fresh until we are ready to eat it.

minerals. These include calcium, magnesium, manganese, phosphorous, and iron. Bananas are nearly fat free and contain no sodium (salt). They are a good source of fiber as well. Fiber is important for digestion.

Bananas have long been a favorite fruit of athletes. That's

because they help to quickly replace essential elements and fluids burned during heavy exercise. Bananas provide fast-acting sugars for quick energy as well. They have sometimes been called a perfect food.

This triathlon participant takes a banana break.

Bananas for

Whether you need a quick breakfast on the run, or have time for a hot-cooked start to your day, bananas make these two recipes nutritious and fun.

Banana Pancakes

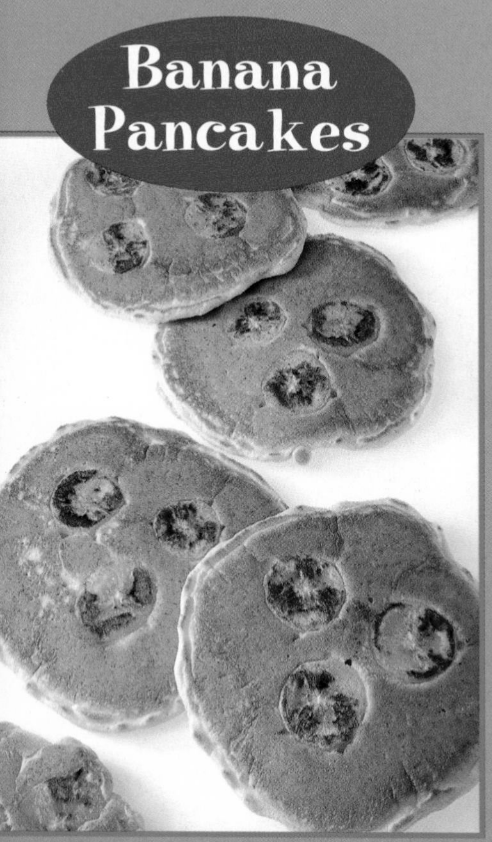

Mix these ingredients until just blended:

1 cup of flour
1 tablespoon of sugar
2 teaspoons of baking powder
1/4 teaspoon of salt
1 cup of milk
1 beaten egg
2 tablespoons of vegetable oil

Coat a large frying pan with cooking spray, and ask an adult to heat the pan until the spray begins to sizzle. Into the pan, pour just under a quarter cup of batter for each pancake. Slice a banana into the pancake when the batter bubbles, just before you are ready to flip the pancake over in the pan. Let this side cook for a minute, then remove from the pan with a spatula. Top with butter or syrup.

Breakfast

In a blender, combine slices of bananas with one or more of your other favorite fruits—strawberries, (fresh or frozen), crushed pineapple, or honeydew melon. Experiment! Add orange juice for a fruity treat, or milk for a creamy one. Blend for half a minute, and pour into a tall glass. Add a few ice cubes, before or after blending, if desired.

Breakfast Shake

Choosing Bananas

At a supermarket, most of the bananas you see are yellow and of average size. These are Cavendish bananas. But there are other types of bananas as well.

Among these are Manzano bananas also known as Apple or Finger bananas. These finger-size bananas turn black after ripening.

(Clockwise from left) Cavendish bananas from Ecuador, Finger bananas and Red Jamaica bananas

Plantains, the largest member of the banana family, are usually cooked before being eaten. Red Jamaica bananas are sweet tasting bananas that turn purplish red when ripe. Apple and Red Jamaica bananas are not as common in the United States as Cavendish bananas. That's because they have thin skins (peels) and bruise easily during shipping.

When selecting Cavendish bananas in the supermarket, here are some important tips to remember:

•The yellower the banana's peel, the sweeter the fruit.

•A banana has completely ripened when brown spots appear on its skin.

•A ripe banana will last longer if kept in the refrigerator.

•To ripen a greenish banana, place it in a brown paper bag overnight. Put an apple or tomato in the bag with it. The natural gases from the fruits helps to ripen both.

•Handle bananas carefully. Dropping or squeezing a banana can bruise it.

Banana Trivia

The banana had a place in art as early as 1500 B.C. Drawings showing bunches of bananas were found on the walls of ancient Egyptian tombs.

Comedians often joke about slippery banana peels. But some people have taken this danger more seriously.

The ancient Egyptians included bananas in their tomb and temple art.

Through the years there have been numerous successful law-suits involving such injuries.

In some tropical countries, people use more than just the banana plant's fruit. They fash-ion baskets, mats, and even

Banana leaf hats (above) and roofs (right) make use of the plant long after its fruit has been picked.

roofs from the leaves of some types of banana plants.

According to The Guiness Book of World Records, the longest banana split ever made was 4.55 miles (7.3

kilometers) long. It was put together by residents of Selinsgrove, Pennsylvania, on April 30, 1988.

When you bite into a banana, you're not just satisfying your hunger. You're eating a healthy food with a fascinating history!

In many parts of the world, bananas are a delicious part of life.

To Find Out More

Here are some additional resources to help you learn more about bananas:

Ancona, George. **Bananas: From Manolo to Margie**. Clarion Press, 1982.

Bloom, Valerie. **Fruits: A Caribbean Counting Poem.** Henry Holt & Co., 1997.

Dowden, Anne Ophelia Todd. **From Flower to Fruit.** Ticknor & Fields, 1994.

Heinrichs, Ann. **Brazil.** Children's Press, 1997.

Powell, Jillian. **Fruit.** Raintree Steck-Vaughn, 1997.

Wexler, Jerome. **Flowers, Fruits, Seeds.** Simon & Schuster, 1987.

Organizations and Online Sites

Bananas on the Web
http://www.geocities.com/NapaValley/1702

Check out the "Encyclopedia Bananica" for all sorts of information and trivia on bananas, in categories that range from show biz and slang to nutrition and literature. Play the Banana Label Game, then click over to the Banana Museum to browse through an amazing collection of "a-peeling stuff."

Dole
651 Ilalo Street
Honolulu, Hawaii 96813
http://www.dole.com/

The fresh fruit giant features a banana page giving tips for selection, storing, and cooking with bananas.

101 Things to Do With a Banana
http://www.dmgi.com/bananas.html

Banana recipes (101 and counting), including cream of banana soup, fried bananas, and banana casseroles, relishes, dumplings, cutlets, and ice cream.

Turbana Corporation
http://www.turbana.com/

This cooperative of banana growers, based in Colombia, presents "The Ripe Stuff"—banana history, recipes, and a ripening guide. Click on the Banana Journey and follow the fruit from plantation to packing plant to pier, bound for U.S. supermarkets.

Important Words

cultivate to plant and tend a crop

debut to present to the public for the first time

export to send goods to other countries

fingers the individual bananas growing on a banana plant

hand the clusters or large bunches of bananas growing on a banana plant

harvest to pick or gather a crop

matures fully develops

pseudostem the trunk portion of a banana plant

peel the outer skin of a banana

Index

(**Boldface** page numbers indicate illustrations.)

Alexander the Great, 11, **11**
Apple bananas, 36–38, **37**
athletes, 32–33, **33**
"banana bags," 26, **26**
banana plants, 7, 21–30, **21**
  fingers of, **23,** 23–24
  hands of, 24
  leaves of, 41–42, **42**
bananas, 20–30
  choosing of, 36–39
  handling of, 39
  history of, 9–19
  peels of, 38, 40–41
  as perfect food, 31–33
  tips for selecting, 38–39
  trivia about, 40–44
banana split, **6,** 7, 42–43
breakfast shake, 35
brown spots, 39
Cavendish bananas, 36–39, **37**
energy, 33
exports, 19
fiber, 32
"fingers," 23–24, **23,** 36, **37**
green bananas, 27–28
greenhouses, 18–19, **19**
Guinness Book of World Records, 42
"hands," 24
harvesting, 8
hothouses, 18–19, **19**
leaves, banana, 41, 42, **42**
Manzano bananas, 36
minerals, 31–32
pancakes, banana 34
paper bag ripening, 39
Philadelphia Centennial Exhibition, 16, **16**
plantains, 37
potassium, 31
pseudostem, 22, **22**
Red Jamaica bananas, 37–38, **37**
refrigeration, 28, 39
ripening, 27–30, **29,** 39
skin color, **32**
sodium, 32
starches, 29
vitamins, 31

Meet the Author

Elaine Landau worked as a newspaper reporter, an editor, and a youth services librarian before becoming a full-time writer. She has written more than one hundred nonfiction books for young people, including True Books on dinosaurs, animals, countries, and food.

Ms. Landau, who has a bachelor's degree in English and journalism from New York University and a master's degree in library and information science from Pratt Institute, lives in Florida with her husband and son.

Photographs ©: Corbis-Bettmann: 16 bottom, 41; ENP Images: 23 right, 26 (Gerry Ellis); Envision: 43 left (Tim Gibson), 2, 43 right (Jack Stein Grove), 25 (Fred Kahn), 42 right (Michael Major), 42 left (J. B. Marshall), cover, 34, 37 right (Steven Needham); Midwestock: 6 right (Michael Rush); New England Stock Photo: 21 left (T. Hannon); North Wind Picture Archives: 10, 11, 14, 16 top; PhotoEdit: 32 right (Robert Ginn), 4, 6 left, 35 (Michael Newman), 32 left (Pat Olear), 37 center (David Young-Wolff); Stock Boston: 33 (Rick Browne), 8 right (Bob Daemmrich), 17 (Robert Fried), 8 left (Martin Rogers), 21 right (Steve Woit); Tony Stone Images: 27 (Mark Joseph), 1 (Victoria Pearson); Visuals Unlimited: 37 left (Walt Anderson), 22 (John D. Cunningham), 23 left (Marc Epstein), 30 (Audrey Gibson), 19 top (N. Pecnik), 19 bottom (Brian Rogers), 29 (Science).